SPACE
A Tour of the Cosmos

D0878701

The Foundations of Science

SPACE
A Tour of the Cosmos

WORKBOOK

TAN Books
Gastonia, North Carolina

Space: A Tour of the Cosmos Workbook © 2022 TAN Books

All rights reserved. With the exception of short excerpts used in critical review, no part of this work may be reproduced, transmitted, or stored in any form whatsoever, without the prior written permission of the publisher.

Unless otherwise noted, Scripture quotations are from the Revised Standard Version of the Bible— Second Catholic Edition (Ignatius Edition), copyright © 2006 National Council of the Churches of Christ in the United States of America. Used by permission. All rights reserved.

Cover & interior design and typesetting by www.davidferrisdesign.com

ISBN: 978-1-5051-2753-9

Published in the United States by
TAN Books
PO Box 269
Gastonia, NC 28053

www.TANBooks.com

Printed in the United States of America

"Where were you when I laid the foundation of the earth?"

−Job 38:4

CONTENTS

A NOTE TO PARENTS

Thank you for using *The Foundations of Science* series to educate your child about God's wonder-filled world of science! Before diving in, make sure to read these brief notes.

WORKBOOKS' PURPOSE

The workbooks in *The Foundations of Science* series are meant to be companions to the texts, simple tools you can use to ensure your child has comprehended the material. But they're also supposed to be fun! The exercises should not feel like a test. Consider letting your students use the text as they answer questions since we just want them to understand the main concepts and remember some of the things they have learned. (We are not trying to stump them!) Younger students especially may need a little "hand-holding" to get some of the answers, but that's okay, and is even encouraged.

TARGET AGE

The workbook is perfect for middle elementary-aged students, but children as young as first grade or as old as fifth can engage with it. There are enough activities that each one does not need to be done and the age of the child can be used to determine which are completed. For example, coloring pages can be used for younger students, while older children may skip those; conversely, younger students may skip some of the personal reflection short answers, while the older ones may be expected to not only answer them but write good and complete sentences. Please cater the workbook to your family's needs.

TYPES OF ACTIVITIES

Most chapters will utilize both a substantive activity (Matching, True/False, Short Answer, etc.), along with something fun, such as a puzzle, word search, coloring page, or arts and crafts. There are also some personal reflection exercises, and most chapters include a question or activity that ties what they studied in that chapter back to the Catholic Faith.

MY SCIENCE JOURNAL

Every chapter begins with a "My Science Journal" spread. Here the students are encouraged to take notes as they read the text, write down questions they have, list the most interesting thing they learned in that chapter, or what they enjoyed the most. They can also log things they saw in nature that week (and it does *not* have to be things that relate to that week's content). This is highly recommended to complete, as it not only helps them comprehend the content better but allows parents to assign a writing exercise as well.

ANSWER KEY

While many of the exercises are subjective and answers will vary, there are also plenty of objective answer exercises that will require grading. An answer key is provided in the back of the book for your use and convenience. If you like, have a conversation about honesty and integrity with your child as you teach them to not peek in the back.

KEY TERMS AND AMAZING FACTS ABOUT SPACE

The "Key Terms and Amazing Facts About Space" included in the text are included here as well. Consider making flash cards for the terms to test your student's knowledge and retention, and let your child sit and relax as they read the facts; seeing them all at once, rather than buried in the text, may help them remember all the fun things they have learned.

WE ARE HERE TO HELP!

We hope we have provided you with everything you need, but if not, don't hesitate to reach out to your friends at TAN Books with any questions you might have.

CHAPTER

1

SPACE
The Galactic Frontier

MY SCIENCE JOURNAL

NOTES:

_____

_____

_____

_____

_____

_____

_____

_____

_____

_____

_____

_____

_____

_____

_____

_____

_____

_____

_____

_____

_____

_____

_____

MY FAVORITE PART OF THIS CHAPTER WAS:

ONE NEW THING I LEARNED WAS:

SOME QUESTIONS I WANT TO ASK ABOUT THIS CHAPTER ARE:

ONE INTERESTING THING I SAW IN NATURE THIS WEEK WAS:

UNDERSTANDING ASTRONOMICAL TERMS

It is important to know the difference between the terms we use to describe different astronomical regions. Write out brief definitions below for the terms based on what you read in the text. Make sure you know the differences between these terms.

Outer Space:

Universe:

Solar System:

Galaxy:

MATCHING

Match the terms with their definitions below.

A. *Density*   **F.** *Cosmology*

B. *Vacuum*   **G.** *Interplanetary space*

C. *Milky Way*   **H.** *Cislunar space*

D. *Interstellar space*   **I.** *Intergalactic space*

E. *Geospace*   **J.** *Observable universe*

1. _____ An area without matter; space is considered this.

2. _____ The scientific study of the creation of the universe.

3. _____ The area of outer space just beyond Earth where satellites orbit.

4. _____ The name for our galaxy.

5. _____ The space between Earth and the Moon.

6. _____ The area of outer space between stars or solar systems.

7. _____ The region of space between galaxies.

8. _____ A measurement of how many particles are packed in a given space.

9. _____ The area of the universe we can detect from our position on Earth.

10. _____ The distance between our star (the Sun) and the planets that orbit it.

FAITH AND SCIENCE

Essay: Our Boundless God

In the space provided, or on a scrap sheet of paper, write a short reflection essay on what the immensity of space teaches us about God. How does meditating on the mysteries and vastness of outer space draw you closer to God? How does it make you love Him more?

_____

_____

_____

_____

_____

_____

_____

_____

_____

_____

_____

_____

_____

_____

_____

_____

_____

_____

_____

_____

_____

_____

_____

_____

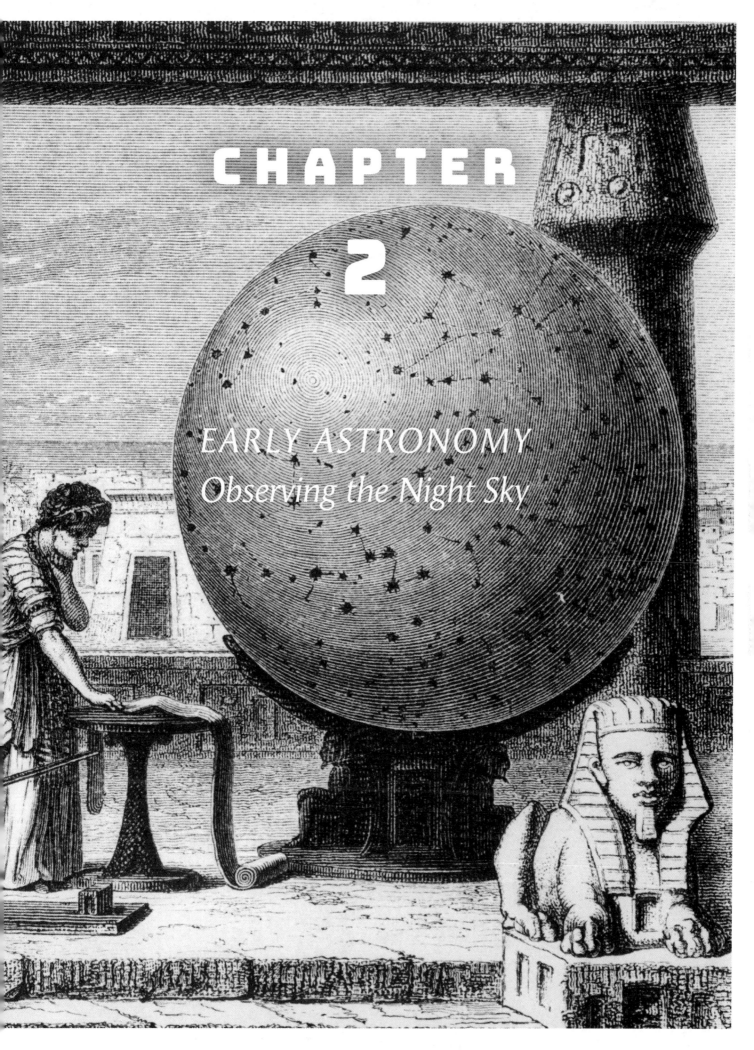

CHAPTER 2

EARLY ASTRONOMY
Observing the Night Sky

MY SCIENCE JOURNAL

NOTES:

_____

_____

_____

_____

_____

_____

_____

_____

_____

_____

_____

_____

_____

_____

_____

_____

_____

_____

_____

_____

_____

MY FAVORITE PART OF THIS CHAPTER WAS:

ONE NEW THING I LEARNED WAS:

SOME QUESTIONS I WANT TO ASK ABOUT THIS CHAPTER ARE:

ONE INTERESTING THING I SAW IN NATURE THIS WEEK WAS:

SHORT ANSWER

1. Explain why some people say astronomy was the "first science"?
What evidence do they provide to back this up?

2. How did early astronomers tell the difference between stars and planets?

3. What is astrology, and why is it not a true science?

4. What is the difference between apparent magnitude and absolute magnitude?

5. What is the difference between a constellation and an asterism?

IDENTIFYING CONSTELLATIONS

With the help of your mom or dad, go online and search for the constellations in the bank below. Then, based on the images below, which depict either the entire constellation or an asterism from it, write the name of each constellation on the line below it.

Ursa Major   Ursa Minor   Cassiopeia   Orion   Gemini   Scorpius

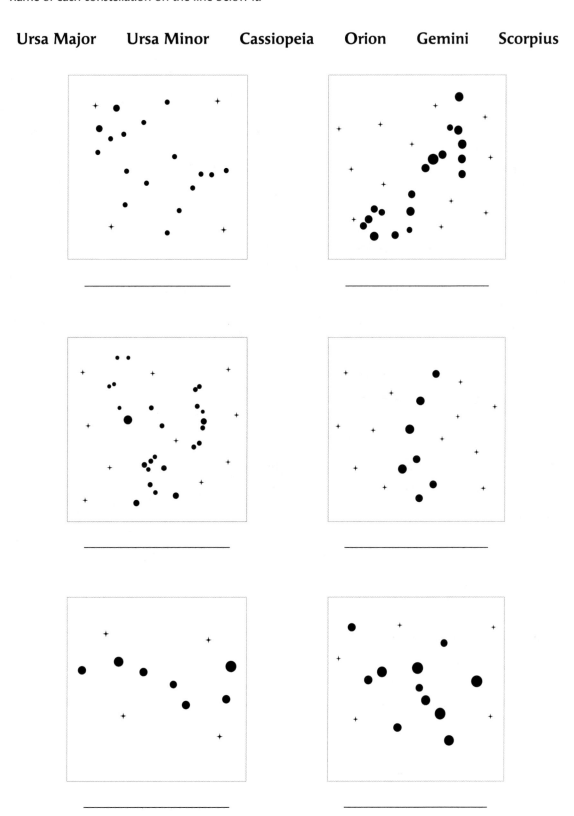

ACTIVITY

Big Dipper Viewer

Materials

1. Paper towel roll tube
2. Foil
3. Rubber band
4. Push pin
5. Scissors

Instructions

1. Cut a 3 x 3 inch square of foil.
2. Place the foil over one end of the tube, centering it.
3. Press or fold the sides of the foil around the tube and secure it with a rubber band or tape.
4. Use the push pin to poke holes into the center of the foil to resemble the Big Dipper.
5. Turn off the lights, go toward a window in the daytime, and, placing the viewer up to your eye, view the Big Dipper!
6. Turn the tube different directions to show that the Big Dipper is not always the same direction in the night sky, due to the rotation of Earth.

Extra Fun! In a completely dark room, you can also shine a flashlight into the tube and project the Big Dipper onto the ceiling.

FAITH AND SCIENCE

Draw Your Own Constellation

Imagine how God must have enjoyed arranging the stars at the creation of the universe. While we don't have the powers He has, it is still fun to create things. Pretend for a moment that God has allowed you to design your own constellation of stars. What would your constellation look like? What would it be called? Consider a constellation that has a connection to the Catholic Faith or some kind of religious symbol. Draw your constellation on the opposite page, and name it at the top.

For fun, see if your mom or dad can figure it out before you draw in the lines and reveal the shape!

CHAPTER
3

THE SUN
Earth's Favorite Star

MY SCIENCE JOURNAL

NOTES:

_____

_____

_____

_____

_____

_____

_____

_____

_____

_____

_____

_____

_____

_____

_____

_____

_____

_____

_____

_____

_____

_____

MY FAVORITE PART OF THIS CHAPTER WAS:

ONE NEW THING I LEARNED WAS:

SOME QUESTIONS I WANT TO ASK ABOUT THIS CHAPTER ARE:

ONE INTERESTING THING I SAW IN NATURE THIS WEEK WAS:

FILL IN THE BLANK

Pick the term from the word key that best completes each sentence.

Gravity   Luminosity
Atoms   Molecular clouds
Element   Stars
Orbit   Red giants
Supernova   Nuclear fusion

1. _____ means an object circles another object, or center of gravity, in a repeatable pattern.

2. The force of attraction that pulls together anything made of matter is _____.

3. While most of space is empty, there are higher concentrated areas full of molecules known as _____, which are known as "stellar nurseries" because they are where stars are born.

4. _____ are astronomical objects, immense spheres made up of superheated gasses.

5. An _____ is pure substance, or form of matter, that consists of only one type of atom.

6. _____ are the basic building blocks of all substances, the tiny units that make up the matter, or physical "stuff" we can see.

7. A measure of a star's brightness, or the amount of energy it emits, is known as its _____.

8. When a star's core runs out of fuel, it collapses and triggers a huge explosion called a _____.

9. The process of _____ combines two (or more) atoms into a single atom; it is the process by which stars put out incredible amounts of energy.

10. After a star has burned through its hydrogen fuel, it starts to go through various changes, like expanding and taking on a reddish color. These stars in the late stage of their evolution are called _____.

UNDERSTANDING MASS

In the box below, explain the difference between mass and weight.

In the box below, explain how we measure the solar mass of stars. What reference point do we use? How does this measurement system work?

A STAR'S COLOR

We learned something interesting about a star's color in this chapter and how it relates to its size, to its lifespan, and how much heat energy it puts off. Do you remember what two colors stars generally are? Using a crayon or marker for these two colors, draw a star below in each box. Then, draw lines from the statements to the correct color.

COLOR 1

Live longer

Live shorter

Hotter/Brighter

Less hot/Less bright

COLOR 2

Bigger

Smaller

FAITH AND SCIENCE

The Stars and the Angels

We talked about stars in this chapter. In the Bible, stars are often seen as symbols of the angels. Color the illustration of the angel below. Then, go find and read Psalm 91:11–12 and write those verses on the lines below the drawing. Have a discussion with your parents about this passage and what you think it means. Then, pray the Guardian Angel Prayer with your family.

Prayer to One's Guardian Angel

Angel of God, my guardian dear,
To whom God's love commits me here,
Ever this day, be at my side,
To light and guard,
To rule and guide.
Amen.

_____

_____

_____

_____

_____

_____

CHAPTER

4

SOLAR SYSTEM TOUR I
Mercury and Venus

MY SCIENCE JOURNAL

NOTES:

_____

_____

_____

_____

_____

_____

_____

_____

_____

_____

_____

_____

_____

_____

_____

_____

_____

_____

_____

_____

MY FAVORITE PART OF THIS CHAPTER WAS:

ONE NEW THING I LEARNED WAS:

SOME QUESTIONS I WANT TO ASK ABOUT THIS CHAPTER ARE:

ONE INTERESTING THING I SAW IN NATURE THIS WEEK WAS:

WHAT IS A PLANET?

What are the three features that define a planet?

1.

2.

3.

Since Pluto only meets the first two of these qualifications,
it is considered a _____ planet.

Mercury: True or False

Write "T" if the statement is true, and "F" is the statement is false.

1. _____ Mercury is the first planet we encounter in our solar system when
moving out from the Sun.

2. _____ Mercury is so small we cannot see it without a telescope.

3. _____ Mercury is blue in color.

4. _____ Mercury is a gas planet.

5. _____ Mercury's days are much hotter than Earth's.

TIME ON MERCURY

In the box below, explain how it is possible for Mercury to have longer days than years.

Venus: True or False

Write "T" if the statement is true, and "F" is the statement is false.

1. _____ Even though Venus is farther from the Sun than Mercury, it is hotter because of its thick atmosphere.

2. _____ The atmosphere on Venus would be friendly to human beings living there.

3. _____ Venus is sometimes called Mercury's "twin."

4. _____ The surface of Venus has many mountains and volcanos.

5. _____ As it orbits the Sun, Venus rotates on its axis in the opposite direction of most planets.

CHAPTER

5

SOLAR SYSTEM TOUR II
Earth, the Moon, and More . . .

MY SCIENCE JOURNAL

NOTES:

_____

_____

_____

_____

_____

_____

_____

_____

_____

_____

_____

_____

_____

_____

_____

_____

_____

_____

_____

_____

_____

MY FAVORITE PART OF THIS CHAPTER WAS:

ONE NEW THING I LEARNED WAS:

SOME QUESTIONS I WANT TO ASK ABOUT THIS CHAPTER ARE:

ONE INTERESTING THING I SAW IN NATURE THIS WEEK WAS:

SHORT ANSWER

1. What do we mean when we say Earth is a "terrestrial planet"?

2. What is the Goldilocks Zone, and why do we call it that?

3. What is a Leap Year, and why do we have it?

4. What do we mean when we say the Moon is "tidally locked" with Earth?

5. What do we mean when we say the Moon does not have its own light?

THE MOON'S PHASES

After studying the chart in your text, come back here and try to label the different phases of the Moon based on the images below.

_____   _____   _____

_____   _____   _____

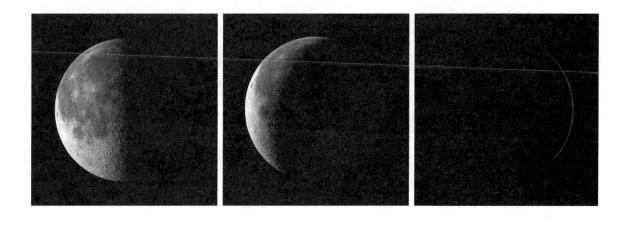

_____   _____   _____

CROSSWORD PUZZLE

Fill in the crossword based on what you read in this chapter.

ACROSS

2. The outer layer of a terrestrial planet

3. A natural satellite

5. The scientific name for the Goldilocks Zone is the _____ habitable zone

6. A New Moon is one of the _____ of the Moon

7. Earth's atmosphere is mostly made up of this gas

9. Earth's core is mostly made up of this metal

DOWN

1. Walked on the surface of the Moon

4. A layer of gas that surrounds a planet or other celestial body

8. Earth is the _____ planet away from the Sun

ACTIVITY

Oreo Moon Phases

Complete this snack project to help you understand the phases of the Moon. (Refer to your text to help make sure you know the different phases and what they look like.)

Ingredients

1. 8 Oreo cookies
2. Cookie sheet pan
3. Large sheet of black construction paper
4. Knife
5. Yellow or white frosting tube

Instructions

1. Lay the construction paper onto the cookie sheet pan, cutting to fit if needed. This will be the night sky.
2. Open each Oreo cookie, carefully leaving the white filling intact on one side.
3. Working from left to right on the top row of the pan, for the New Moon, place one plain non-filled side of a cookie on the upper left side of the pan. (*Note: if you want them to stick you can use the frosting, placing down dabs of it like glue.*)
4. For the Waxing Crescent, use the knife to carve out the Oreo filling into a thumbnail shape so the filling is on the outer right side of cookie.
5. For the First Quarter Moon, use the knife to carve out half of the Oreo filling so the filling is on the right side of cookie.
6. For the Waxing Gibbous, use the knife to carve out only about 1/4 of the Oreo filling, leaving about 3/4 of the filling on the right side of cookie.
7. Now working on the bottom row of the sheet, place an Oreo down with the full filling for the Full Moon.
8. Next comes the Waxing Gibbous, where 3/4 of the filling is now on the left side of the cookie.
9. Then comes the Second Quarter Moon, where half of the cookie is white again, but this time the filling is on the left side of the cookie.
10. Finally, for the Waxing Crescent, carve out almost 3/4 of the white Oreo filling to leave only 1/4 of it showing on the left side of the cookie.
11. Label each moon phase using the writing frosting.
12. Display and enjoy (and eventually eat with some milk)!

For extra fun, look out into the night sky when you do this activity and see what current phase the Moon is in!

CHAPTER

6

SOLAR SYSTEM TOUR III
Mars

MY SCIENCE JOURNAL

NOTES:

_____
_____
_____
_____
_____
_____
_____
_____
_____
_____
_____
_____
_____
_____
_____
_____
_____
_____
_____
_____
_____
_____
_____

MY FAVORITE PART OF THIS CHAPTER WAS:

ONE NEW THING I LEARNED WAS:

SOME QUESTIONS I WANT TO ASK ABOUT THIS CHAPTER ARE:

ONE INTERESTING THING I SAW IN NATURE THIS WEEK WAS:

UNDERSTANDING ORBITS

In the box below, explain what we mean when we say that all the planets in our solar system orbit the Sun on a plane. Why does this happen?

What do the terms *perihelion* and *aphelion* refer to in regards to a planet's orbit?
Answer in the box below.

MULTIPLE CHOICE
Choose the best answer.

1. What color is Mars?
   A. Blue
   B. Whitish/Yellow
   C. Red

2. Mars is a:
   A. terrestrial planet.
   B. gas giant.
   C. star.

3. *True or False?* Mars is smaller than Earth.
   A. True
   B. False

4. Mars has the _____ yet discovered in our solar system.
   A. deepest canyon
   B. tallest mountain
   C. largest moon

5. Most of Mars's atmosphere (over 90 percent) is made up of:
   A. carbon dioxide.
   B. oxygen.
   C. nitric oxide.

LIFE ON MARS?

We often speculate if there could be life on other planets. Mars is our closest neighbor, so it is fitting to wonder if life could exist there. In the box below, list as many challenges as you can remember from the text that would make it very unlikely for life to exist on Mars.

MARS WORD SEARCH

Find the following words from the chapter in the word scramble below.

Mars	**perihelion**
red planet	**aphelion**
rust	**orbit**
iron oxide	**Phobos**
Mons Olympus	**Deimos**

```
A M G P S R Q B E U Q F G S T
P J E R E O E D E I M O S N B
H X A N R R I I W Y P F R V Z
E M L L A X I A M J I H U T F
L A F O O L S H V O B T S G E
I X W N F Y P D E T S X T J O
O H O O N Z S Y R L P A F D G
N R J Z E Y Q S E C I L P S E
I B L X L E Q A H F F O Y C T
M O N S O L Y M P U S I N I G
T P H O B O S D D E R A B M L
A B A E M F I U I A Z R Y V E
L R E D P L A N E T O G X A I
O B V J I F V X Q F V E U O F
D L G V F T I A J N J Z K T W
```

CHAPTER

7

SOLAR SYSTEM TOUR IV
Jupiter and Saturn

MY SCIENCE JOURNAL

NOTES:

_____
_____
_____
_____
_____
_____
_____
_____
_____
_____
_____
_____
_____
_____
_____
_____
_____
_____
_____
_____
_____

MY FAVORITE PART OF THIS CHAPTER WAS:

ONE NEW THING I LEARNED WAS:

SOME QUESTIONS I WANT TO ASK ABOUT THIS CHAPTER ARE:

ONE INTERESTING THING I SAW IN NATURE THIS WEEK WAS:

FACTS ABOUT JUPITER

In the box below, list as many things as you can remember about the planet Jupiter that you learned in the text.

In the box below, explain what we mean when we say Jupiter does not technically orbit the Sun like Earth and other planets do.

FACTS ABOUT SATURN

In the box below, list as many things as you can remember about the planet Saturn
that you learned in the text.

In the box below, describe what the famous rings of Saturn are like.

FILL IN THE BLANK: JUPITER AND SATURN

Complete each sentence with the right word or term. Use your text if you need help.

1. Jupiter is the _____ planet in our solar system.

2. Jupiter and Saturn are both _____ rather than rocky, terrestrial planets; this means they are sometimes called _____.

3. Jupiter has what is called the _____, which is a massive storm bigger than Earth that continually churns in the atmosphere. A similar large storm on Saturn is called the _____.

4. Both Jupiter and Saturn have many _____.

5. It is possible that it may rain _____ on both Jupiter and Saturn!

WHAT'S IN A NAME?

Explain the meaning and origin of each planet's name (this was found in one of the inserts between chapters in the text).

Jupiter:

_____

_____

_____

Saturn:

_____

_____

_____

Coloring Fun!

Find images of Jupiter and Saturn online and color the illustrations below based on how they look.

CHAPTER

8

SOLAR SYSTEM TOUR V
Uranus and Neptune

MY SCIENCE JOURNAL

NOTES:

_____
_____
_____
_____
_____
_____
_____
_____
_____
_____
_____
_____
_____
_____
_____
_____
_____
_____
_____
_____
_____
_____
_____

MY FAVORITE PART OF THIS CHAPTER WAS:

ONE NEW THING I LEARNED WAS:

SOME QUESTIONS I WANT TO ASK ABOUT THIS CHAPTER ARE:

ONE INTERESTING THING I SAW IN NATURE THIS WEEK WAS:

PUT THE PLANETS IN ORDER

Now that we have been introduced to all the planets in our solar system, see if you can name them all and put them in order, identifying them on the diagram below.

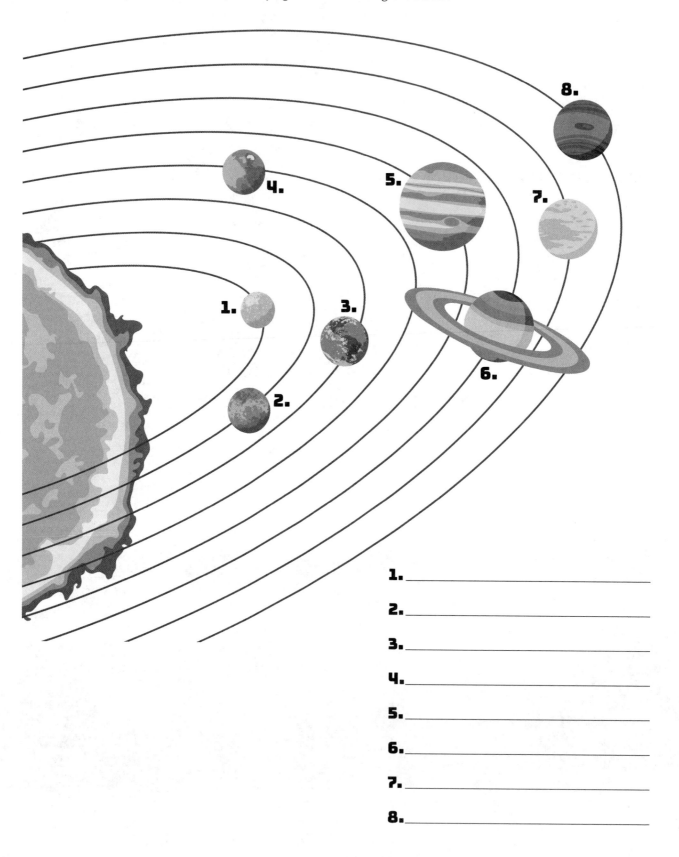

1._____

2._____

3._____

4._____

5._____

6._____

7._____

8._____

CROSSWORD PUZZLE

Fill in the crossword based on what you read in this chapter.

ACROSS

3. A large storm on Neptune that was once seen but has since disappeared.

5. Both Uranus and Neptune are considered this.

6. Like Saturn, Uranus has _____.

7. Uranus has many _____.

9. Because of its extreme tilt and long orbit, Uranus has long _____.

10. The color of Uranus.

DOWN

1. Largest moon of Neptune.

2. Neptune is about _____ times larger than Earth.

4. Neptune's _____ orbit makes it unique from other planets.

8. NASA mission that helped us learn about planets in the distance reaches of the solar system.

ACTIVITY

Scaling Down the Solar System

To help get a sense for how far each planet is away from the Sun, at least on a smaller scale, complete the following activity.

The table below has an approximation of the (average) distance from the Sun to all of the planets, scaled down so that, for example, the distance from Neptune to the Sun is one hundred feet.

Start by finding a long sidewalk or empty parking lot and use chalk to mark out the distances, or use cones or flags in a big open field. Set your starting point first. This is where you will mark the position of the Sun. Measure one hundred feet in a straight line—the one-hundred-foot mark is where you should place Neptune. After that, use the table below with the distances provided to mark the other planets. Also included is about how big each planet would be on this scale (but essentially, they would all just be really tiny). The Sun is massive, but would only be four inches when scaled down to match the distances—approximately the size of an orange! None of the planets would be larger than a ladybug on your map of the solar system. In fact, Jupiter would only be the size of a flea!

	Sun	Mercury	Venus	Earth	Mars	Jupiter	Saturn	Uranus	Neptune
Distance from Sun	0	36 million	67 million	93 million	142 million	484 million	886 million	1.784 billion	2.793 billion
Scaled distance	0' 0"	1' 3.5"	2' 5"	3' 4"	5' 1"	17' 4"	31' 8.5"	63' 10.5"	100'
Scaled size	4"	0.0013"	0.0032"	0.0034"	0.0018"	0.038"	0.032"	0.014"	0.013"

Note: In the table, ' = feet and " = inches. So 1' 3.5" = 1 foot and 3.5 inches.

ACTIVITY

Planetary Mnemonic

Do you know what a "mnemonic" is? It's when you use a pattern of letters or words to help you remember the order of something. So you might come up with a sentence where each word starts with a certain letter that follows the same order as what you are trying to remember. "Please Excuse My Dear Aunt Sally" is a famous one used in math. (Ask your parents about it if you have never heard of it.)

Come up with a mnemonic to help you remember the order of the planets. The sentence you come up with must have words that start with the below letters, since the planets are ordered: Mercury, Venus, Earth, Mars, Jupiter, Saturn, Uranus, Neptune. An example is provided below to help you. Remember, it doesn't have to make sense, it just needs to help you remember the order of the planets. Extra points for strange and goofy ones!

Example: My Very Energetic Mother Just Slurped Up Nuggets

M

V

E

M

J

S

U

N

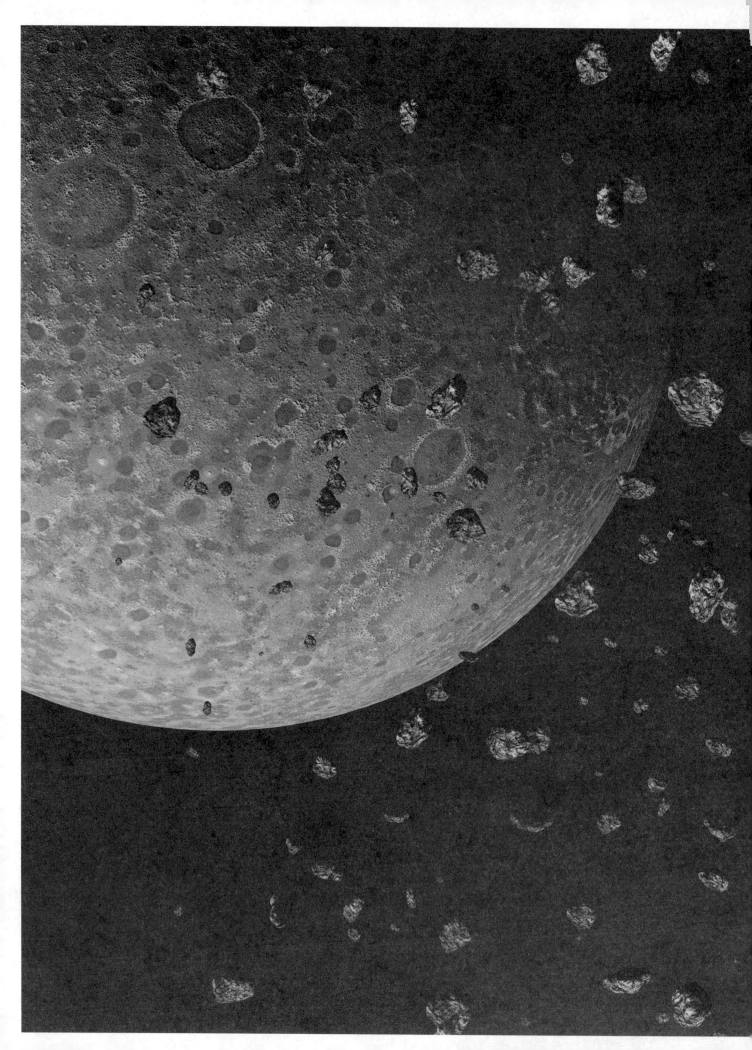

CHAPTER
9

SOLAR SYSTEM TOUR VI
Pluto, Dwarf Planets, Asteroids, and More!

MY SCIENCE JOURNAL

NOTES:

_____
_____
_____
_____
_____
_____
_____
_____
_____
_____
_____
_____
_____
_____
_____
_____
_____
_____
_____
_____
_____
_____
_____

MY FAVORITE PART OF THIS CHAPTER WAS:

ONE NEW THING I LEARNED WAS:

SOME QUESTIONS I WANT TO ASK ABOUT THIS CHAPTER ARE:

ONE INTERESTING THING I SAW IN NATURE THIS WEEK WAS:

Pluto: True or False

Write "T" if the statement is true, and "F" is the statement is false.

1. _____ Pluto has more moons than Earth.

2. _____ Pluto is a planet.

3. _____ Pluto is always farther out than all the planets.

4. _____ Pluto is made up primarily of rock and ice.

5. _____ Pluto is located in the Kuiper Belt.

On the lines below, explain the difference between a Meteoroid, a Meteor, and a Meteorite.

Meteoroid:

_____

_____

Meteor:

_____

_____

_____

Meteorite:

_____

_____

_____

MATCHING

Match the terms with their definitions below.

A. *Kuiper Belt*   **F.** *Asteroid*

B. *Eris*   **G.** *Meteor*

C. *Trojan asteroid*   **H.** *Comet*

D. *Halley's Comet*   **I.** *Ceres*

E. *Dwarf Planet*   **J.** *Meteorite*

1. _____ When a meteoroid enters our atmosphere, or another planet's, it typically burns up as this.

2. _____ Means "little star."

3. _____ Pluto is an example of this.

4. _____ Formed by ice and dust, can have a tail when they approach the Sun.

5. _____ Largest object in the asteroid belt.

6. _____ Visible from Earth once every seventy-five years.

7. _____ Asteroids that travel along the same orbit as a planet.

8. _____ A region in our solar system where many things orbit that are smaller than a planet.

9. _____ Largest dwarf planet in our solar system by mass.

10. _____ Any piece of a meteor that makes it through to land on a planet's surface.

ACTIVITY

The Solar System in 3D

Materials

1. Variety pack of various sized Styrofoam balls
2. 8 wooden skewers (extra for backup)
3. Various colors of spray paint or craft paint
4. Large red plastic cup
5. Large paper plate
6. Scissors

Instructions

1. Construct the planets as follows for the size of the Styrofoam ball and paint colors. (The ball sizes do not have to be exactly as listed below.)

   - SUN: 5-inch ball, bright yellow
   - MERCURY: 1 ¼-inch ball, orange
   - VENUS: 1 ½-inch ball, blue-green
   - EARTH: 1½-inch ball, dark blue with green highlights
   - MARS: 1 ¼-inch ball, red
   - JUPITER: 4-inch ball, orange with red and white stripes.
   - SATURN: 3-inch ball, yellow-orange (for the rings, you can construct a donut shaped circle out of a paper plate.
   - URANUS: 2-inch ball, cobalt blue
   - NEPTUNE: 2 ½-inch ball, light blue

2. After the Sun has dried, place it into the plastic cup, which is used as the "base" of the project. (Or, the Sun can just rest on a table or the ground.)
3. After the planets have dried, place a skewer in each.
4. Using your knowledge of the distance of each planet from the Sun, cut the skewers to resemble the proper distance. Mercury, being the shortest distance, would have the shortest skewer. Neptune, being the farthest distance, would have the longest skewer.
5. Stick the skewers into the Sun, all around it, so they appear to be orbiting the Sun. Make sure the planets are evenly spread out around the Sun for balance!
6. Spin your sun and watch the planets orbit it!

MAZE

We learned in the text that Halley's Comet comes close enough to Earth for us to see it every 75 years. Complete the maze to bring this famous comet closer to our home planet!

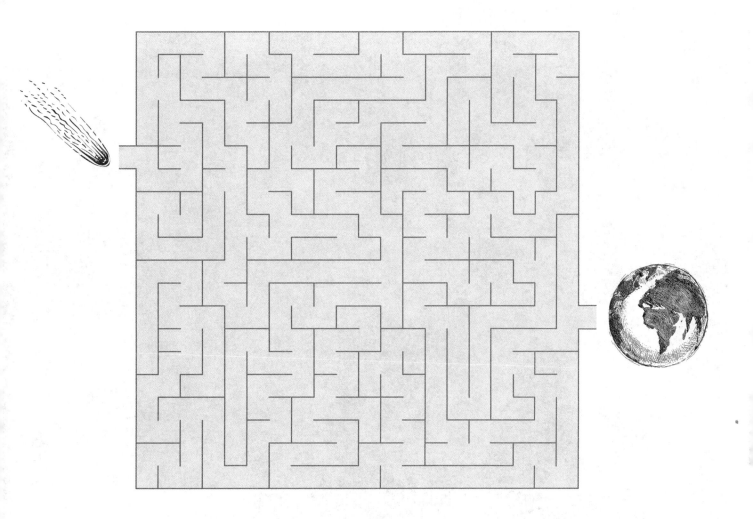

CHAPTER

10

TELESCOPES
*Exploring the Solar System
from Earth and in Space*

MY SCIENCE JOURNAL

NOTES:

_____
_____
_____
_____
_____
_____
_____
_____
_____
_____
_____
_____
_____
_____
_____
_____
_____
_____
_____
_____
_____
_____
_____

MY FAVORITE PART OF THIS CHAPTER WAS:

ONE NEW THING I LEARNED WAS:

SOME QUESTIONS I WANT TO ASK ABOUT THIS CHAPTER ARE:

ONE INTERESTING THING I SAW IN NATURE THIS WEEK WAS:

MULTIPLE CHOICE
Choose the best answer.

1. Who was the Italian scientist who improved on the design of early telescopes to make them more powerful?

    A. Johannes Kepler

    B. Neil Armstrong

    C. Galileo Galilei

2. The heliocentric model states that:

    A. the Sun is the center of our solar system.

    B. Earth is the center of our solar system.

    C. the Sun is the center of the Milky Way.

3. Which of the following makes it difficult to see into space even with telescopes?

    A. clouds

    B. human activity

    C. Earth's magnetic field

4. To see into the distance reaches of space, we put high-powered telescopes:

    A. on top of mountains.

    B. up in space.

    C. All of the above.

5. This was launched in 1990 and was for many years the most powerful and versatile telescope in operation:

    A. Hubble Space Telescope

    B. Halley Space Telescope

    C. Vatican Advanced Technology Telescope (VATT)

TELESCOPES WORD SEARCH

Find the following words from the chapter in the word scramble below.

Telescope	Galileo
Hubble	Kepler
Observatories	Vatican
Heliocentric	Nebula

```
F V M R H V N N H I K P F U C
A O A C Z C E J X S H K N Q H
E I B T X X B D P K B K J T E
W C Z S I N U T X M C J V L L
X U T K E C L O M R N Q P E I
G T U X D R A H P N F S P J O
X L L S D Z V N R H J O V F C
K E P L E R Z A V Y C I C E E
R C B H A U O W T S G N B P N
E L Q N U E A B E O T E N B T
B G S N L C L L Z L R U F D R
F B X I K L E R A L Y I N X I
U F L X V T S T Z U N B E O C
K A K E Q W I A F N F D O S Z
G B H K H U B B L E R V H T G
```

ACTIVITY

Understanding How Telescopes Work

Supplies Needed

1. Glass cup filled with water
2. Pad or piece of paper and pen

Not all of us have access to a telescope, but we can do simple experiments to show how they work. One is to fill a clear glass full of water. Take a piece of paper and write your name on it, not too big, not too small. Then, hold the paper up, and hold the glass of water before it, closing one eye to view it better, focusing in on your name. Do you see how it is slightly larger while looking through the water? Why is this?

Well, remember how telescopes make use of curved lenses and light rays to make distant objects appear nearer. In a similar way, light rays bend when they pass through water, thus enlarging what we see through it. Of course, telescopes magnify things to a much greater degree, but this neat example can give you an idea of how the science behind it works.

LEARNING MORE ABOUT THE VATICAN'S EFFORTS
TO EXPLORE SPACE

By using both the textbook and online sources (with the supervision of your parents), write down five things the Catholic Church is currently doing or has done in the past to further humanity's understanding of outer space.

1.

2.

3.

4.

5.

CHAPTER

11

SPACE EXPLORATION
Astronauts, Shuttles, and Stations

MY SCIENCE JOURNAL

NOTES:

MY FAVORITE PART OF THIS CHAPTER WAS:

ONE NEW THING I LEARNED WAS:

SOME QUESTIONS I WANT TO ASK ABOUT THIS CHAPTER ARE:

ONE INTERESTING THING I SAW IN NATURE THIS WEEK WAS:

SHORT ANSWER

1. What does NASA stand for?

2. What was the "Space Race"?

3. What was Project Apollo? What was its primary goal?

4. Who were the first two astronauts to land on the Moon?

5. What are "unmanned missions," and why are they beneficial to our exploration of outer space?

ESSAY

In the space provided, or on a scrap sheet of paper, choose one of the two topics below and write a brief essay.

1. Do outside research on the International Space Station. Write enough about its history and current efforts that you can teach a family member about it. Read your essay to them.

2. Do outside research on a famous astronaut, from the past or today, of your choosing. Talk about his or her life and what accomplishments he or she achieved to help further humanity's goal of exploring space. Read your essay to a family member.

continue on next page

FAITH AND SCIENCE

Saint Joseph Cupertino Prayer Card

Saint Joseph of Cupertino is the patron saint of astronauts because it is said that during his life he would sometimes levitate during Mass or when praying the Liturgy of the Hours. Other stories describe him bounding forward, as if unconstrained by gravity, perhaps just like astronauts on the Moon. This is why Saint Joseph is often depicted flying through the air in artwork (though, of course, not in a space suit).

In the template below, draw and color a picture of Saint Joseph flying through the stars, then write a prayer to him on the back praying for all astronauts.

ACTIVITY

Bottle Rocket Experiment

Complete only with adult supervision!

Materials

1. Small plastic water bottle, emptied (16.9 fl oz)
2. 2–3 cups vinegar
3. 3 Tbsp baking soda
4. Cork that will fit securely onto the bottle
5. 3 wooden pencils of the same size, or dowels the size of a pencil
6. Lightweight tape such as clear packaging tape
7. 3 x 3 inch (approximate) paper towel square
8. Safety goggles

Instructions

1. Remove the lid from the water bottle.

2. Turning the bottle upside down, tape the three pencils around (eraser side down) evenly so that the pencils can hold up the bottle. (Try not to use so much tape that it weighs it down.)

3. Set the bottle aside.

4. Create a "fuel packet" by filling a 3 x 3 inch square of paper towel with about 3 tbsp baking soda, rolling it up to create a snug packet so none spills out.

5. Go outside and find a clear, safe space to launch your rocket.

6. When ready, fill the bottle about 1/3 to 1/2 full of vinegar.

7. Put on your goggles.

8. With the rocket set before you, the cork in one hand, and the fuel packet of baking soda in the other, be ready to launch.

9. Quickly drop the fuel packet into the bottle, place the cork in (not too tightly), turn the bottle over and run away to safety.

10. The rocket should launch into the air. If not, check your amounts of baking soda and vinegar combination or adjust the tightness of the cork.

CHAPTER

12

DISTANT SPACE
Galaxies and Exoplanets

MY SCIENCE JOURNAL

NOTES:
_____
_____
_____
_____
_____
_____
_____
_____
_____
_____
_____
_____
_____
_____
_____
_____
_____
_____
_____
_____
_____
_____

MY FAVORITE PART OF THIS CHAPTER WAS:

ONE NEW THING I LEARNED WAS:

SOME QUESTIONS I WANT TO ASK ABOUT THIS CHAPTER ARE:

ONE INTERESTING THING I SAW IN NATURE THIS WEEK WAS:

DEFINE THE TERMS

Write out a definition for each term based on what you read in the text.

Exoplanet:

_____

_____

_____

_____

Star systems:

_____

_____

_____

_____

Galaxy:

_____

_____

_____

_____

Black hole:

_____

_____

_____

_____

Light-year:

_____

_____

_____

_____

ACTIVITY

Understanding the Transit Method

Do you remember what the transit method was? It's a way we can detect exoplanets that orbit other stars and know how big they are. When the exoplanet passes between us and the star it orbits, we can note how the star's luminosity dims slightly. In some ways, it's like seeing a shadowed "ball" pass over the bright star. This lets us know that planet is there.

To get a sense of what this might look like, complete the following brief activity.

Supplies

1. Flashlight

2. Tennis ball or some other similar ball

3. String (several feet)

Directions

1. Go into a dark room and sit before an open space on the wall.

2. Turn your flashlight on and shine it on the wall from a few feet back so that it creates a yellow-white circle on the wall. This will be your sun or star.

3. Have a family member tie the string around the ball so that they can hang it between the flashlight and the wall, moving it from left to right or right to left like a planet in orbit.

4. Note the way the shadow passes before the reflection on the wall. This is similar to how the transit method works!

FINAL ESSAY

In the space provided, or on a scrap sheet of paper, pick one topic that you learned about in this unit on space to teach someone else about. Write about this topic and why you found it so interesting. It could be one of the planets, or asteroids, or black holes, or something from the history of astronomy. Write down as much as you can remember!

_____

_____

_____

_____

_____

_____

_____

_____

_____

_____

_____

_____

_____

_____

_____

_____

_____

_____

_____

_____

_____

_____

_____

_____

AMAZING FACTS ABOUT SPACE

- Even if you could travel at the speed of light (which you really couldn't because your body couldn't withstand the force), it would still take you *billions of years* to travel from the center of the observable universe to the outer edge!

- Outer space, the area *between* all the objects in our universe, is almost entirely empty. There are so few molecules in space, or such a low enough density of particles, that space is considered a vacuum. This means that unlike our atmosphere on Earth, there isn't air to breathe in outer space since there are no molecules like oxygen or carbon dioxide floating around. This is why astronauts have to wear their suits and helmets when they travel into outer space, because they receive oxygen through them.

- The distance between Earth and the Moon is about 239,000 miles!

- The distance between the Sun and the distant planet Neptune is around 2.8 billion miles! That's over eleven thousand trips between Earth and the Moon.

- The Big Bang Theory proposes that the universe started in one intense explosion (a "big bang") about fourteen billion years ago.

- Molecular clouds are also called "stellar nurseries" because these regions of space are where new stars are "born."

- The largest stars are bluish in color, burn hotter and brighter, and have shorter lifespans than smaller stars. Meanwhile, the smallest stars live longer, have lower temperatures, and look reddish orange. Put another way, larger stars have more fuel (hydrogen to convert into helium) but use it up at a much faster rate, which results in more energy being released, which gives us a brighter star.

- Most stars in the universe that we can observe are smaller stars—more than 70 percent of stars are in the smallest size class.

- If you were to travel straight through it, our sun is around 864,000 miles across. This makes it a middle- to small-sized star.

- The core of the Sun (the center), where fusion happens and energy originates, is twenty-seven million degrees Fahrenheit! The surface of the Sun, meanwhile, is around ten thousand degrees. Surprisingly, it takes more than ten thousand years for this energy to escape from the Sun's core to be released outwards. From there, light energy would travel ninety-three million miles through space to reach Earth. Though that might sound like a long distance—and it is—it can complete this trip in about eight minutes!

- Our Sun has enough fuel to last several more billion years.

- When the core of a star runs out of fuel, it collapses and triggers a huge explosion called a supernova. Scientists can study the remnants of supernovas to learn more about them. There is one in the constellation of Cassiopeia, and it shows gases and material speeding away from the explosion (which happened hundreds of years ago) at more than thirty million miles an hour! This sends elements out throughout the galaxy, including to Earth!

- Mercury is the smallest planet of the eight "official" planets in our solar system and is about eighteen times smaller than Earth. This means Mercury is much closer to the Moon in its size than it is to Earth.

- When you look up at night, some of the brighter things that look like stars might actually be planets rather than stars.

- Mercury's surface temperature can reach over eight hundred degrees Fahrenheit!

- Since Mercury has a small orbit around the Sun (88 "Earth days") but spins (rotates) very slowly (sun-up to sun-down 176 "Earth days"), its days are longer than its years.

- Even though Venus is farther from the Sun than Mercury, Venus is the hottest planet in the solar system because it has a thick atmosphere with greenhouse gases, while Mercury has no atmosphere. Temperatures can reach almost nine hundred degrees Fahrenheit on Venus!

- Earth and Venus are sometimes referred to as twin planets because they are similar in size, mass, density, and gravity (though Venus is a little bit smaller).

- Venus has mountains and volcanos on it. Several of the volcanos are thought to be still active, while its tallest mountain is over 7 miles high. By way of comparison, Mount Everest, Earth's tallest mountain, is only 5.5 miles high!

- Like Mercury, day to night cycles on Venus are longer than the time it takes to orbit the Sun (days and nights are longer than Venus "years"). It takes about 225 Earth days for Venus to complete its trip around the Sun, but Venus's days are about 243 Earth days.

- Venus orbits the Sun in the same direction as other planets, but its rotation is opposite of most planets, so if you could stand on Venus, you would observe the Sun rise in the west and set in the east.

- On average, Earth is about ninety-three million miles away from the Sun. (We say "on average" because our orbit isn't perfectly round, so it is not always the same distance away.) This distance keeps our planet from becoming too hot, while the atmosphere keeps it from becoming too cold.

- The air temperature on Earth, measured near the surface, ranges from a minimum of -128 degrees Fahrenheit (in Antarctica) to around 134 degrees Fahrenheit (measured in Death Valley, California). The average global temperature is a much more tolerable 60 degrees.

- Though our calendars have 365 days in a year, it actually takes the Earth 365 days, 5 hours, 59 minutes, and 16 seconds to travel around the Sun. This is approximately an extra quarter of a day longer than our calendars. To compensate for this, we have what are known as "Leap Years," where an extra day is added in February every four years to get us back on track.

- Though Earth's days are twenty-four hours—equal to 86,400 seconds—over the past century, Earth has slowed down almost imperceptibly due to various factors such as movement in the Earth's core, tectonic movements, and earthquakes, making days a few milliseconds longer. So, as with leap years, when we add an extra day, people have added a "leap second" from time to time to make up for the difference. The last leap second occurred on December 31, 2016. All at the same time, clocks worldwide added an extra second, counting all the way to sixty instead of fifty-nine (the minute changing at sixty-one).

- Earth's moon is tidally locked, or has a "synchronous rotation." This means the Moon rotates on its axis in the same amount of time that it takes it to travel around Earth so that one side, or face, of the Moon is constantly facing Earth. Once a celestial body starts orbiting another, the gravity between the objects acts to slowly decrease the rotation until it becomes "locked." This is where the phrase "far side of the Moon" comes from. In order to see this mysterious far side of the Moon, we either need to send satellites into space to observe it, take a spacecraft there ourselves, or else create a time machine to go back to the time before the Moon was locked with the one side always facing Earth!

- Mars orbits the Sun at a distance of around 142 million miles, or almost 50 million miles farther away from the Sun, on average, than Earth.

- Mars is about half the size of Earth, making it the second smallest planet in the solar system. (It would be the third smallest if Pluto were still recognized as a planet.)

- Even though Mars is smaller than Earth, it has a taller mountain and deeper canyon than our planet. Olympus Mons is an ancient volcano that forms the tallest mountain on Mars. It stands at almost fifteen miles high! That makes it about three times larger than Mount Everest, our tallest mountain. Meanwhile, Mars's deepest canyons reach four miles into the ground. For reference, the Grand Canyon is only about one mile deep!

- Overall, Mars is much colder than Earth, with an average temperature of about negative eighty degrees Fahrenheit. Like Earth, though, there is quite a lot of temperature variation across the planet. Mars is hotter at its equator, up to seventy degrees, and much colder at its poles, dropping below negative two hundred degrees.

- Since Mars takes almost twice as long to orbit the Sun (687 Earth days), Martian seasons are much longer than Earth's.

- Jupiter is 484 million miles from the Sun, more than three times the distance from the Sun to Mars!

- Jupiter is the largest planet in the solar system, measuring 86,881 miles in diameter (the width of the planet, or distance through the center). To help put its size into context, Jupiter's diameter is about *eleven times bigger* than Earth. Jupiter is so big that if you combined the mass of all the other planets in our solar system, this "mega-planet" would still be smaller than Jupiter!

- Jupiter rotates faster than any other planet in the solar system, completing one day (one rotation around) once every nine hours and fifty-five minutes. One trip around the Sun for Jupiter, however, takes about twelve Earth years.

- One of the most famous features of Jupiter is its Great Red Spot. This "spot" is larger than Earth and is actually an ongoing gigantic storm, a slow-rotating cyclone (it takes six days to complete one cycle) detectable by early telescopes dating back all the way to 1665. This storm, and others on Jupiter's surface, are caused, in part, by the rapid movement of the planet and the turbulent movement of its atmosphere.

- A current count of confirmed moons puts Jupiter with seventy-nine moons and Saturn with eighty-two, but we are still discovering more small moons around these planets (and Jupiter may have more than one hundred).

- Saturn is the second largest planet in the solar system, only about 80 percent of the size of Jupiter in diameter (Jupiter has much more mass, though). Saturn is still much larger than Earth (almost ten times bigger in diameter).

- Saturn's rings appear at around four thousand miles away from the surface, and the rings are about seventy miles wide! The rings' depth varies, but are about seventy feet deep, on average. Most of the particles that make up this seventy-foot-high wall of ice and dust are small, but some larger chunks of ice and debris exist too, with some as large as thirty feet. Since they could be observed from Earth with early telescopes, Saturn's rings were described as far back as the 1600s.

- Saturn only takes about 10.5 hours to spin around once, making a "day" on Saturn less than half as long as days on Earth. Similar to Jupiter, though, it takes the planet much longer than Earth to orbit the Sun. Saturn makes one trip around the Sun about once every twenty-nine years.

- One fascinating feature of Jupiter and Saturn is that it may actually rain diamonds during storms on these planets! Diamonds are formed from the element carbon through intense heat and pressure. Carbon is on these planets in the form of methane, and the high heat and pressure contained on the planet may turn this carbon into diamonds.

- Uranus and Neptune are the farthest planets from the Sun. Uranus is 1.784 billion miles away, and Neptune is 2.793! It takes Uranus about 84 Earth years to orbit the Sun, while Neptune takes 165 years!

- Like Earth, Uranus rotates on a tilt (giving it a wobble as it spins), but the tilt is almost ninety degrees. This extreme tilt means that one pole is completely pointed toward the Sun, while the other is in total darkness. As it orbits, these poles slowly swap places, but this "winter" at each pole lasts for twenty years because of the planet's long orbit (eighty-four years).

- When we view Uranus through a high-powered telescope, it appears blue because of the methane. The methane traps red light but lets bluish light bounce off and return to Earth.

- Neptune's largest moon is named Triton. It is unique in that it orbits opposite Neptune's spin (called a retrograde orbit). All of Neptune's other moons orbit in the same direction of the spin of the planet. In fact, almost all other moons in the solar system, including Earth's moon, orbits its planet in the direction of the planet's spin (the few exceptions being some tiny, distant moons of the gas giants).

- Neptune is the only planet in the solar system that is not visible from Earth without a telescope. Interestingly, it was discovered by math. Uranus was discovered by Sir William Herschel in 1781 when he was searching for comets. As astronomers continued to study this newly discovered planet, they noticed that something large was tugging on the planet with its gravity. Based on the observed effects on Uranus from an unknown source, astronomers were able to calculate not only where to find Neptune in the night sky but also approximately how large the planet was.

- Pluto, a dwarf planet, is smaller than Earth's moon, measuring 1,477 miles across in the middle. Over 70 percent of Pluto is its core. Interestingly, scientists estimate that there could be a liquid-water ocean surrounding this massive core, an ocean that may be as much as one hundred miles deep!

- Pluto is so far out in the solar system that it takes 248 years to orbit the Sun and can get as low as negative four hundred degrees Fahrenheit!

- A common misconception is that asteroid belts (or fields), if we could fly through them, would be difficult to navigate, such a place where you may bounce your spacecraft off any number of large, rocky objects. In reality, objects in the asteroid belt are millions of miles apart, on average. Instead of being difficult to fly through this area of space, it's actually the opposite; it is quite difficult to get a satellite or space probe to pass close enough to an object you want to study.

- Ceres is the largest object in the asteroid belt of our solar system. It's a dwarf planet about 587 miles across, or a little larger than the size of Texas.

- Asteroid means "little star"; this makes sense when you consider they appear as small points of light when viewed from a telescope. But rather than balls of hot gases, like a star, asteroids are small, rocky objects.

- A meteoroid is similar to an asteroid but smaller. These objects in space are small and rocky, ranging from grains of rice (or smaller) up to the size of asteroids. When a meteoroid enters our atmosphere, or another planet's, it typically burns up—then it's known as a meteor. Any piece that makes it through to land on the planet is a meteorite.

- Halley's Comet is visible from Earth (without a telescope!) around once every seventy-five years. This is because the comet's orbit takes about seventy-five years, and it is only visible as it gets close to Earth. It travels out past Neptune at the distant part of its orbit but then ends up closer to the Sun than Earth when at its nearest point. It was last visible from Earth in 1986, and so will not return until 2061!

- Some people say that astronomy was the first science because we have archeological evidence of ancient people recording their observations of the night sky.

- When early astronomers tracked some of the planets as bright lights in the sky, they counted Venus as two different stars. It is most visible in the late evening and early morning, so it was called the Morning and Evening Star.

- Some of the shapes in the night sky you may recognize, or could look for on a clear night, are not technically constellations. These can be referred to as asterisms. For example, the well-known Big Dipper is not actually a constellation, as most people call it. Instead, in more technical terms, it is an asterism.

- The earliest clear record of telescope technology came from the very early 1600s, when various Dutch spectacle-makers and scientists each tried to patent the device in 1608. An Italian scientist named Galileo Galilei heard of these devices and set out to improve the telescope design. In just a short time, he increased the power of his telescopes; early versions could only increase images 3x in size, but soon he had a telescope with 23x magnification!

- The Vatican Observatory has been in operation, supported by the Holy See, since 1774, when it was established as the Observatory of the Roman College. Currently, the Vatican Observatory conducts observations in Arizona (though it still has headquarters in Italy).

- Electric lights and other light from human activity makes it more difficult to see the stars. The night sky is "washed out" by this light pollution. This makes it easier to see many stars out in the country rather than a city. It is also a main reason why the Vatican uses a mountain in Arizona for observations.

- Astronomers as early as the 1800s realized that there would be advantages to making observations of space *from* space rather than somewhere on Earth. But it wasn't until the late 1960s and early 1970s that we were able to put operational telescopes into space. The first space telescope was the Orbiting Astronomical Observatory 2 (OAO-2, also nicknamed the Stargazer), sent into orbit by the United States in 1968.

- Planetary nebulas are expanding shells of gas released by red giants. They are sometimes referred to as the "eyes of God" because of their round, almost eyelike appearance.

- One of the most fascinating observations the Hubble Telescope revealed are the Pillars of Creation, which appears to be an image of stars in an area known as the Eagle Nebula, found on another arm of our own Milky Way galaxy. These giant cloud-like structures are estimated to be ~5,700 light-years away from us, and each pillar to be several light-years high. For scale, our entire solar system is less than 2 light-years across!

- The first object sent into space was a V-2 rocket launched from Germany in 1944. The United States captured V-2 rockets, improved the technology, and used them to take photos of Earth in 1946 (from sixty-five miles in the air), to send fruit flies into space in 1947 (the first animals), and to send a rhesus monkey named Albert II to space in 1949. These tests allowed us to better understand possible effects on living things so we could know whether humans could safely travel through space.

- The word *astronaut* comes from the Greek *astron*, meaning "star," and *nautes*, meaning "sailor." So while a more technical definition of astronauts would call them people who are trained and equipped to travel into outer space, it sounds a lot more fun to call them "star sailors!"

- On July 16, 1969, a Saturn V rocket carried the Apollo spacecraft into space. There were three astronauts on board—Neil Armstrong, Buzz Aldrin, and Michael Collins. Neil Armstrong and Buzz Aldrin were the first human beings to walk on another celestial body. (Michael Collins stayed in orbit around the Moon as pilot of the spacecraft they used for the return home.)

- People from at least nineteen different countries have visited or lived on the International Space Station for a period of time, and five space agencies help operate the station: NASA (United States), Roscosmos (Russia), CSA (Canada), ESA (Europe), and JAXA (Japan).

- The Milky Way has as many as four hundred billion stars and is estimated to be at least one hundred thousand light-years across! It is thought to be one of billions of galaxies—a 2021 estimate put the number at two hundred billion—but there may even be 1 trillion!

- There are more stars in the sky than there are grains of sand on all the beaches of Earth.

- Gravity is so strong near black holes that nothing can escape from it—not spaceships, planets, particles of dust, or even light escapes.

- Even though Proxima Centauri is the second closest star to Earth after our own sun, it is almost 4.3 light-years away.

- Kepler-16b is an exoplanet found in the constellation of Cygnus that orbits two suns. This means if you could stand on the surface (unlikely since it's a gas giant), you would have two shadows.

NASA SPACE PROGRAM HISTORY

One-man
Mercury Spacecraft

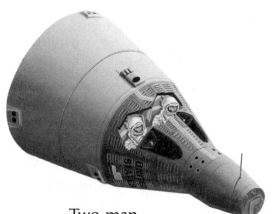

Two-man
Gemini Spacecraft

Three-man
Apollo Spacecraft

Program	Start date	First crewed flight	End date	Number of crewed missions launched	Notes
Mercury Program	1958	1961	1963	6	First U.S. crewed program
Gemini Program	1961	1965	1966	10	Program used to practice space rendezvous and EVAs
Apollo Program	1960	1968	1972	11	Landed first humans on the Moon
Skylab	1964	1973	1974	3	First American space station
Apollo–Soyuz Test Project	1971	1975	1975	1	Joint with Soviet Union
Space Shuttle program	1972	1981	2011	135	First missions in which a spacecraft was reused
Shuttle-Mir program	1993	1995	1998	11	Russian partnership
International Space Station	1993	1998	Ongoing	65	Joint with Roscosmos, CSA, ESA, and JAXA; Americans flew on Russian Soyuz after 2011 retirement of Space Shuttle
Commercial Crew Program	2011	2020	Ongoing	3	Current program to shuttle Americans to the ISS
Artemis program	2017	Ongoing	Ongoing	0	Current program to bring humans to the Moon again

KEY TERMS

Absolute magnitude – *Chapter 2*: How bright an object is; this is opposed to how bright it seems to us when we look at it (known as *apparent magnitude*).

Aphelion – *Chapter 6*: The point in the orbit of a planet, asteroid, or comet in which it is farthest from the Sun.

Apparent magnitude – *Chapter 2*: How bright an object appears to an observer on Earth; this is opposed to how bright it actually is (*absolute magnitude*).

Asterisms – *Chapter 2*: A prominent pattern or group of stars within a constellation.

Asteroid – *Chapter 9*: Small, rocky objects hurtling through outer space, often orbiting a star; *asteroid* means "little star."

Astronomy – *Introduction*: The branch of science which deals with celestial objects, space, and the physical universe.

Atmosphere – *Chapter 5*: A layer of gas that surrounds a planet or another celestial body.

Atoms – *Chapter 3*: The basic building blocks of all matter, made up of protons, neutrons, and electrons.

Big Bang Theory – *Chapter 1*: A theory that proposes the universe started in one intense explosion (a "big bang") about fourteen billion years ago.

Black hole – *Chapter 12*: A location in space with immense mass, which generates intense gravitational forces; gravity is so strong near black holes that nothing can escape from it.

Circumstellar habitable zone – *Chapter 5*: Also known as the "Goldilocks Zone," this is the region in our solar system where the right distance from the Sun provides temperatures that makes life possible; *circumstellar* means in orbit around a star, and *habitable* means that life may be able to exist.

Cislunar space – *Chapter 1*: The region of space between Earth and the Moon (about 238,900 miles).

Comet – *Chapter 9*: Celestial objects formed of ice and dust; can manifest a "tail" when close enough to the Sun, formed by ice that vaporizes from the heat of the Sun, and then streams outward (the "tail" points away from the Sun).

Constellation – *Chapter 2*: A region of space in the night sky where a collection of bright stars outline a particular shape or pattern.

Cosmology – *Chapter 1*: The area of science that studies how the universe came into existence and how it develops or changes over time.

Cosmos – *Introduction*: The ordered physical universe that surrounds us, governed by specific laws (physics, gravity, etc.).

Crust – *Chapter 5*: The outer layer of a terrestrial planet.

Density – *Chapter 1*: A measurement of how many molecules are packed in a per unit area. (The formula would be density = mass/volume, or mass contained in a certain amount of space.)

Dwarf planet – *Chapter 4*: See *Planet*.

Eclipse – *Chapter 6*: When two celestial bodies line up with the Sun (or star/light source), the middle celestial body will block light from the Sun, or cast a shadow on the outside celestial body. A *lunar eclipse* occurs when Earth falls between the Moon and the Sun in a precise way so that Earth's shadow is cast on the Moon, while a *solar eclipse* occurs when the Moon passes between Earth and the Sun; finally, if the Moon passes into just the right position to block the whole of the Sun, we call that a *total solar eclipse.*

Element – *Chapter 3*: Pure substance, or form of matter, that consists of only one type of atom; iron, oxygen, carbon, and gold are all examples of elements.

Exoplanet – *Chapter 12*: Any planet outside our solar system, meaning it orbits a different star.

Galaxy – *Chapters 1 & 12*: A huge collection of stars, or solar systems, gravitationally bound to one another.

Gas Giant – *Chapter 7*: A large planet (Jupiter and Saturn) made up primarily of hydrogen and helium, much like stars; in fact, they are sometimes referred to as "failed stars" because they are similar in makeup to stars but are not as large and do not conduct nuclear fusion to generate energy.

Geospace – *Chapter 1*: The region of space nearest Earth, where satellites orbit.

Gravity – *Chapter 3*: The force that keeps our feet planted firmly on the ground; a force of attraction that pulls objects together—or things made of matter (technically speaking, anything that has mass or energy). The larger and denser the object, the stronger the gravitational pull.

Halley's Comet – *Chapter 9*: A famous comet visible from Earth (without a telescope!) around once every seventy-five years because of its orbit.

Heliocentric model – *Chapter 10*: The theory that states that the Sun is the center of the solar system rather than Earth; *helio*=sun, *centric*=center. Collected evidence consistently and clearly supports this idea, which is why we can call it a theory.

Hubble Space Telescope – *Chapter 10*: Hubble for short, this was the most powerful and versatile space telescope in operation until recently, launched into orbit in 1990 aboard the space shuttle Discovery; continues to this day to make observations.

Hypothesis – *Chapter 1*: A statement or proposed explanation for an observation found in a nature.

Ice giant – *Chapter 8*: Planets like Uranus and Neptune with a high temperature and high pressure "ocean" of water and liquid below clouds.

Intergalactic space – *Chapter 1*: The region of space between galaxies.

International Space Station – *Chapter 11*: A large artificial satellite orbiting Earth since 1998 where astronauts from over a dozen countries have lived and worked; five space agencies help operate the station: NASA (United States), Roscosmos (Russia), CSA (Canada), ESA (Europe), and JAXA (Japan).

Interstellar space – *Chapter 1*: The region between stars or solar systems.

Interplanetary space – *Chapter 1*: The region of space within our solar system between our star (the Sun) and the planets that orbit it.

Kuiper belt – *Chapter 9*: A region in our solar system spanning just outside Neptune's orbit where many objects smaller than a planet (dwarf planets, asteroids, etc.) orbit the Sun, including Pluto; it orbits the Sun in roughly the same plane as the planets, which makes it sort of like a flat, wide belt encircling the solar system.

Light-year – *Chapter 4 & 10*: The distance that light travels in one year, an astonishing six trillion miles (~5.88 trillion, to be more precise).

Luminosity – *Chapter 3*: In regards to stars, a measure of the energy they emit; we see this as brightness.

Mass – *Chapter 3*: A measurement of the amount of matter an object or a physical body contains.

Matter – *Chapter 1*: The physical "stuff" we can see, like planets, plants, animals, etc. contained in the universe, including human beings.

Meteor – *Chapter 9*: A meteoroid that burns up as it enters Earth's atmosphere (or any planet's).

Meteorite – *Chapter 9*: Any piece of a meteor that makes it through Earth's atmosphere (or any planet) to land on the planet.

Meteoroid – *Chapter 9*: Small, rocky objects hurtling through outer space, like asteroids but smaller, ranging from the size of grains of rice (or smaller) up to the size of asteroids; becomes a meteor if it enters Earth's atmosphere.

Molecular cloud – *Chapter 3*: An area of space where there is a higher concentration of molecules, places where dust and gas has accumulated over time to reach a higher density; they are also called "stellar nurseries" because these regions of space are where stars are born.

Moon – *Chapter 5*: A natural satellite—an object that orbits a planet or other celestial body.

Nuclear fusion – *Chapter 3*: A process whereby stars forge or fuse multiple atoms into a single atom; much of the fusion that occurs within stars is the fusion of hydrogen atoms, with their single proton, into one atom of helium, which has two protons.

Observable universe – *Chapter 1*: The region of the universe we can detect, or observe, from our position on Earth.

Observatories – *Chapter 10*: Places on Earth where large telescopes are housed, giving astronomers special tools to study the cosmos.

Orbit – *Chapter 3*: When an object circles another object, or center of gravity, in a repeatable pattern.

Orbiting Astronomical Observatory 2 (OAO-2) – *Chapter 10*: Nicknamed "Stargazer," was the first telescope to launch into outer space, sent into orbit by the United States in 1968.

Outer space (space) – *Chapter 1*: The area beyond Earth that stretches between our home planet and all the other astronomical objects (planets, moons, stars, galaxies, etc.); outer space is the empty area between all the objects in our universe.

Parallax – *Chapter 3*: The observed movement of an object by changing the position, or point of view, of the observer.

Perihelion – *Chapter 6*: The point in the orbit of a planet, asteroid, or comet in which it is closest to the Sun.

Photon – *Chapter 3*: A particle of light or electromagnetic radiation.

Planet – *Chapter 4*: The largest objects in our solar system that orbit the Sun; for an object to be considered a planet, it must meet three criteria: (1) orbit a star (so for our solar system, they must orbit the Sun); (2) be large enough that it has enough gravity to force it into a nearly round (spherical) shape; (3) be large enough that its gravity clears away objects of similar size near its orbit around the Sun. If something meets the first two criteria but not the last, it is considered a *dwarf planet*.

Planetary nebula – *Chapter 10*: An expanding shell of gas released by red giants; sometimes referred to as the "eyes of God" because of their resemblance to eyeballs.

Probe – *Chapter 11*: Unmanned spacecraft designed to transmit information about the environment they are sent in to.

Project Apollo – *Chapter 11*: The NASA program launched in the late 1960s with the goal of landing human beings on the Moon.

Red giant – *Chapter 3*: Stars that are in the late stages of their evolutionary "life"; they take on a reddish color as they burn through their hydrogen fuel.

Rocket – *Chapter 11*: A projectile craft that moves itself by sending exhaust, from combustion or a chemical reaction, in the opposite direction of its movement.

Star systems – *Chapter 12*: Groups of stars close enough to each other to affect one another with their gravity; in effect, they all orbit each other, bound by each other's gravity.

Space – See *Outer Space*.

Space Race – *Chapter 11*: The name given to the competition between the United States and the Soviet Union (or USSR) as they each tried during the 1950s and '60s to achieve better space-flight technology and capability than the other.

Solar system – *Chapter 1*: The gravitational area around our sun.

Star – *Chapter 3*: An astronomical object, an immense sphere made up of superheated gases (or more technically, plasma—you could read more about this on your own); our sun is a star.

Supernova – *Chapter 3*: The explosion that is the result of a star collapsing ("dying").

Syzygy – *Chapter 6*: When three celestial bodies align in a row (ex. Earth, Moon, Sun).

Telescope – *Chapter 10*: An important instrument astronomers use to make observations; uses curved lenses and light rays to make distant objects (planets, stars, etc.) appear nearer.

Terrestrial plants – *Chapter 5*: Planets like Mercury, Venus, Earth, and Mars that are composed of rocky matter (land).

Theory – *Chapter 1*: An explanation for something found in nature based on evidence and repeatedly tested ideas. In science, this is distinct from a hypothesis, prediction, or idea because a theory is an explanation that has been repeatedly tested and supported by a broad range of evidence.

Tidally locked – *Chapter 5*: A phenomenon by which an orbiting astronomical body always has the same face toward the object it is orbiting, as in the case of the Earth's moon.

Transit method – *Chapter 12*: A method of detecting the presence, size, and composition of celestial bodies (i.e., exoplanets) whereby the object passes between an observer (us) and a larger celestial body (the star it orbits), dimming the star and thus providing information about the object in question.

Trojan asteroids – *Chapter 9*: Asteroids that travel along the same orbit as a planet, either traveling in front of or trailing behind the planet in its orbit.

Universe – *Chapter 1*: All of the existing space, time, energy, and matter considered as a whole.

Unmanned missions – *Chapter 11*: Voyages into outer space with no human being aboard the spacecraft.

Vacuum – *Chapter 1*: An area without matter (the Latin root means "vacant" or "void"); space is a vacuum.

Vatican Advanced Technology Telescope (VATT) – *Chapter 10*: A special, powerful telescope run by the Vatican Observatory and housed in Arizona.

Near side of the Moon

Far side of the Moon

ANSWER KEY

For Parents

CHAPTER 1

UNDERSTANDING ASTRONOMICAL TERMS

It is important to know the difference between the terms we use to describe different astronomical regions. Write out brief definitions below for the terms based on what you read in the text. Make sure you know the differences between these terms.

1. **Outer Space** — This is the area beyond Earth that stretches between our home planet and all the other astronomical objects (planets, moons, stars, galaxies, etc.). It is essentially empty, the space *between* all the objects in the universe.

2. **Universe** — This term includes all of space and the time, energy, and matter contained there.

3. **Solar System** — This term refers to the region around a given star and everything that orbits it.

4. **Galaxy** — This term refers to a huge collection of solar systems (or stars). Our galaxy is called the Milky Way.

MATCHING

1. B
2. F
3. E
4. C
5. H
6. D
7. I
8. A
9. J
10. G

CHAPTER 2

SHORT ANSWER

1. Explain why some people say astronomy was the "first science"?
What evidence do they provide to back this up?

Some people say that astronomy was the first science because we have archeological evidence of ancient people recording their observations of the night sky. A few of these earliest observations are cave drawings which appear to be records of important or impressive astronomical events, like supernovas or eclipses. Some of the very first written historical records we have discovered are those of Babylonians, from several thousand years ago, making observations of planets, stars, and astronomical events. But this interest in astronomy is also almost universal across all cultures; there are examples like this from China, various parts of Europe, the Americas, and more.

2. How did early astronomers tell the difference between stars and planets?

They were able to tell the difference because the stars were stationary in the sky while the planets moved.

3. What is astrology, and why is it not a true science?

Astrology uses astronomical observations to try to predict events that affect people. This is not a science (and is even heretical) because clearly the movement of stars and planets does not give us predictions of the future. Instead, we know the movements of the planets in the sky are simply determined by their set path, or orbit, in relation to ours. Furthermore, the relative brightness of the planets doesn't indicate their importance but instead that they are quite close to us. It is heretical to practice astrology because God's providence is what oversees the unfolding of the future.

4. What is the difference between apparent magnitude and absolute magnitude?

The brightness of an object as it appears to an observer on Earth is its apparent magnitude (apparent -> how it appears to be). This is different from the measurement of how bright the object actually is—this measurement is called the absolute magnitude.

5. What is the difference between a constellation and an asterism?

A constellation is a region of space in the night sky where a collection of bright stars outline a particular shape or pattern. An asterism, meanwhile, is a smaller collection of stars within a constellation.

IDENTIFYING CONSTELLATIONS

Gemini

Scorpius

Orion

Ursa Minor

Ursa Major

Cassiopeia

CHAPTER 3

FILL IN THE BLANK

1. **Orbit**

2. **Gravity**

3. **Molecular clouds**

4. **Stars**

5. **Element**

6. **Atoms**

7. **Luminosity**

8. **Supernova**

9. **Nuclear fusion**

10. **Red giants**

UNDERSTANDING MASS

In the box below, explain the difference between mass and weight.

Mass is a measurement of an object, sort of like weight, but not really the same. The mass of an object is a measurement of the amount of matter it contains (atoms, molecules, etc.). You might think of it as how much space an object takes up. Weight, meanwhile, is how much force gravity pulls on a particular object, given its mass. On Earth, our weight is the same as our mass because we define these measurements based on Earth's gravity. But on the Moon, our mass would be the same (mass is *always* the same), but we would *weigh* less because the Moon doesn't have as much gravitational pull.

In the box below, explain how we measure the solar mass of stars. What reference point do we use? How does this measurement system work?

We measure the mass of stars based on a comparison of it to our own star, the Sun. If a star is smaller than the Sun, this value would be less than one. If a star has ten times more mass than the Sun, the value would be ten, meaning that it takes ten suns put together to get the same mass.

A STAR'S COLOR

Note: Your child could draw either color in either box. Simply make sure the lines are going to the right color.

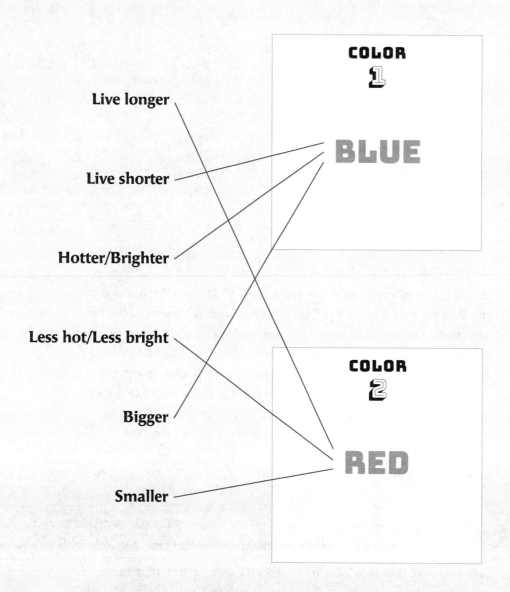

COLOR
1

BLUE

Live longer

Live shorter

Hotter/Brighter

Less hot/Less bright

COLOR
2

Bigger

RED

Smaller

CHAPTER 4

WHAT IS A PLANET?

What are the three features that define a planet?

1. **Orbit a star (so for our solar system, they must orbit the Sun).**

2. **Be large enough (enough mass) that its own gravity pulls itself into a nearly round (spherical) shape.**

3. **Be large enough (enough mass) that it clears away objects of similar size near its orbit around the Sun.**

Since Pluto only meets the first two of these qualifications,

it is considered a _____ **dwarf** _____ planet.

MERCURY: TRUE OR FALSE?

1. **T**

2. **F, it can be seen with the naked eye!**

3. **F, gray, similar to our moon.**

4. **F, it is made of rock and metals, like Earth.**

5. **T**

TIME ON MERCURY

In the box below, explain how it is possible for Mercury to have longer days than years.

Mercury completes one rotation on its axis in 59 Earth days. But since Mercury is also moving around the Sun during this time, it's actually about 176 Earth days between Mercury's sunrise and sunset. So one "day" on Mercury would feel like six months to us. And while Earth orbits the Sun in 365 days (giving us our year), since Mercury is closer to the Sun and its orbit is shorter, a "year" on Mercury is only 88 days. Thus, a Mercury year takes less time than a Mercury day!

VENUS: TRUE OR FALSE?

1. T

2. F, quite the opposite!

3. F, Venus is called Earth's twin.

4. T

5. T

CHAPTER 5

SHORT ANSWER

1. What do we mean when we say Earth is a "terrestrial planet"?

   This means Earth is made up of rocky matter, or what you might think of as the land we walk on every day (as opposed to planets composed mostly of gas).

2. What is the Goldilocks Zone, and why do we call it that?

   The Goldilocks Zone is scientifically known as the circumstellar habitable zone. It refers to the region surrounding a star (our sun) where life is possible because the temperature is just right, not too cold, not too hot, like Goldilocks says in the fairy tale.

3. What is a Leap Year, and why do we have it?

   We say that it takes Earth 365 days to orbit the Sun, giving us our year. But it actually takes 365 days, 5 hours, 59 minutes, and 16 seconds. This is approximately an extra quarter of a day longer than our calendars. To account for this, every four years we have a "Leap Year", where our calendars have 29 days in February instead of 28. Essentially, we save up that extra quarter of a day for four years, and then spend them on an extra day in February to get back on track.

4. What do we mean when we say the Moon is "tidally locked" with Earth?

   The Moon has a "synchronous rotation" with Earth. This means it rotates on its axis in the same amount of time that it takes it to travel around Earth so that one side, or face, of the Moon is constantly facing Earth. Once a celestial body starts orbiting another, the gravity between the objects acts to slowly decrease the rotation until it becomes "locked"; thus, we say it is "tidally locked."

5. What do we mean when we say the Moon does not have its own light?

   The light we see when we look at the Moon does not emanate from the Moon itself; it is a reflection of the light of the Sun hitting. Therefore, we say the Moon does not have its own light.

THE MOON'S PHASES

New Moon

Waxing Crescent

First Quarter

Waxing Gibbous

Full Moon

Waning Gibbous

Last Quarter

Waning Crescent

New Moon

116

CROSSWORD PUZZLE

CHAPTER 6

UNDERSTANDING ORBITS

In the box below, explain what we mean when we say that all the planets in our solar system orbit the Sun on a plane. Why does this happen?

A plane refers to a flat surface. If we drew an imaginary, flat surface around the Sun, each planet moves in a circle on roughly that same surface (the same plane).

It turns out that disks or planes around various celestial bodies are common in astronomy. As these objects form, dust and matter accumulate around them. Balances between forces like gravity (pulling things inward) and centrifugal force (forcing orbiting objects outward) acting on these objects tend to collect everything together onto the same plane as they spin around.

What do the terms *perihelion* and *aphelion* refer to in regards to a planet's orbit? Answer in the box below.

These two terms refer to the points at which a planet is the farthest away and the closest to the sun during its orbit.

Perihelion refers to the closest point; aphelion refers to the farthest.

MULTIPLE CHOICE

1. C
2. A
3. A
4. B
5. A

LIFE ON MARS?

We often speculate if there could be life on other planets. Mars is our closest neighbor, so it is fitting to wonder if life could exist there. In the box below, list as many challenges as you can remember from the text that would make it very unlikely for life to exist on Mars.

- **Very little oxygen**
- **No liquid water**
- **Subfreezing temps in most places**
- **UV radiation**

MARS WORD SEARCH

CHAPTER 7

FACTS ABOUT JUPITER

- **Jupiter is the largest planet in the solar system; its diameter is about *eleven times bigger* than Earth.**

- **The immense size of Jupiter makes it the only planet that doesn't technically orbit the Sun but rather it orbits a point between it and the Sun.**

- **Jupiter is a gas giant rather than a terrestrial (or rocky) planet.**

- **Jupiter is also the most "extreme" planet. In terms of spin, the planet rotates faster than any other planet in the solar system, completing one day (one rotation around) once every nine hours and fifty-five minutes.**

- One trip around the Sun for Jupiter takes about twelve Earth years.

- Jupiter's "Great Red Spot" is a storm larger than Earth constantly swarming in the atmosphere. This storm, and others on Jupiter's surface, are caused, in part, by the rapid movement of the planet and the turbulent movement of its atmosphere.

- Jupiter has at least seventy-nine moons but there are possibly more yet undiscovered. Not all of Jupiter's moons are small. In fact, some are quite large, even larger than some planets! Jupiter's largest moon, Ganymede, is larger than Mercury.

- More facts can be found in the text.

In the box below, explain what we mean when we say Jupiter does not technically orbit the Sun like Earth and other planets do.

The immense size of Jupiter makes it the only planet that doesn't technically orbit the sun. Essentially, when one object orbits another, the smaller one doesn't exactly circle around the larger object. Instead, both objects have gravity, *pulling on each other*, and both objects orbit their combined center of gravity.

If two things are the same size, the center of gravity would be the middle point between them—think of two people of the same size sitting on a seesaw. As one object increases in size, the center of gravity shifts towards the larger object. The center of gravity between Earth and a tiny satellite orbiting it, for example, is almost exactly in the center of Earth. This is because Earth is so much more massive than the satellite. Because all the other planets are so small compared to the Sun, the center of gravity between them is some-where in the middle of the Sun. This means the smaller planets circle around the Sun with basically no effect on it. But Jupiter is so large that the center of gravity between it and the Sun lies about thirty thousand miles above the surface of the Sun—thus, both objects technically orbit that point in space between them.

FACTS ABOUT SATURN

- Like Jupiter, Saturn is a gas giant. It is the second largest planet in the solar system, only about 80 percent of the size of Jupiter, but still much larger than Earth (almost ten times bigger).

- Scientists' current thinking is that the core of Saturn is composed of iron and nickel.

- Layers of metallic hydrogen on Saturn give it a strong magnetic field, much stronger than that of Earth.

- The most famous feature of Saturn is its distinctive rings. Images of Saturn typically show these rings orbiting the planet like a flat disc away from its equator. These rings are primarily made up of ice, but also include small amounts of dust and rocky material. Saturn's rings were described as far back as the 1600s.

- Saturn only takes about 10.5 hours to spin around once, making a "day" on Saturn less than half as long as days on Earth. Similar to Jupiter though, it takes the planet much longer than Earth to orbit the Sun. Saturn makes one trip around the Sun about once every twenty-nine Earth years.

- Saturn generates heat of its own, almost like a star. Despite this internal energy, the temperature in the clouds of Saturn is about –288 degrees Fahrenheit. The planet is so far from the Sun—886 million miles away!—that it doesn't receive as much solar energy as planets nearer the Sun, like Earth.

- Saturn has periodic giant storms that form in its clouds, which may be similar to the Great Red Spot on Jupiter (though not as permanent). A "Great White Spot" can be observed once per Saturn year.

- More facts can be found in the text.

In the box below, describe what the famous rings of Saturn are like.

The most famous feature of Saturn is its distinctive rings. Images of Saturn typically show these rings orbiting the planet like a flat disc away from its equator. These rings are primarily made up of ice, but also include small amounts of dust and rocky material. Rings appear at around four thousand miles away from the surface, and the rings are about seventy thousand miles wide! The rings' depth varies, but are about seventy feet deep, on average. Most of the particles that make up this seventy-foot-high wall of ice and dust are small, but some larger chunks of ice and debris exist too, with some as large as thirty feet. Since they could be observed from Earth with early telescopes, Saturn's rings were described as far back as the 1600s.

FILL IN THE BLANK: JUPITER AND SATURN

1. largest
2. gas giants; failed stars
3. Great Red Spot; Great White Spot
4. moons
5. diamonds

WHAT'S IN A NAME?

1. Jupiter (Jove) was the king of the Roman gods and was the god of the sky and thunder. This is fitting because Jupiter is the biggest planet in our solar system (and it has the Great Red Spot raging on it night and day).

2. Saturn (Saturnus) was the Roman god of agriculture and wealth, and was Jupiter's father.

CHAPTER 8

PUT THE PLANETS IN ORDER

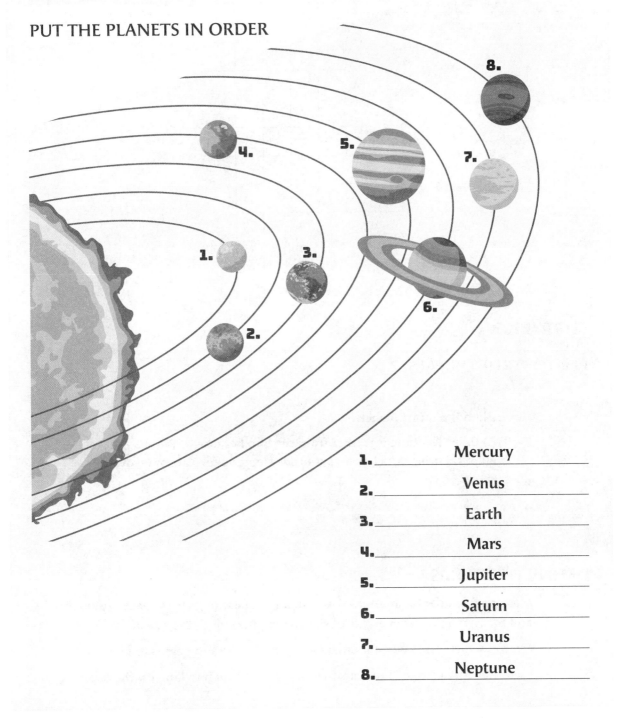

1. _____ Mercury

2. _____ Venus

3. _____ Earth

4. _____ Mars

5. _____ Jupiter

6. _____ Saturn

7. _____ Uranus

8. _____ Neptune

CROSSWORD PUZZLE

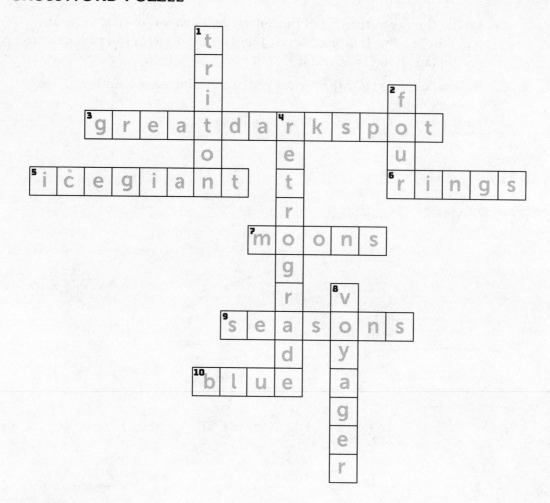

CHAPTER 9

PLUTO: TRUE OR FALSE?

1. T
2. F, Pluto is a dwarf planet.
3. F, At times Neptune is farther from the Sun because of Pluto's exaggerated elliptical orbit, which brings it *inside* Neptune for part of its orbit
4. T
5. T

DEFINE THE TERMS

Meteoroid: **Similar to an asteroid, but smaller. These objects in space are small and rocky, ranging from grains of rice (or smaller) up to the size of asteroids.**

Meteor: **A meteoroid when it enters our atmosphere and begins to burn up.**

Meteorite: **Any piece of a meteor that makes it through to land on the planet.**

MATCHING

1.	G	6.	D
2.	F	7.	C
3.	E	8.	A
4.	H	9.	B
5.	I	10.	J

MAZE

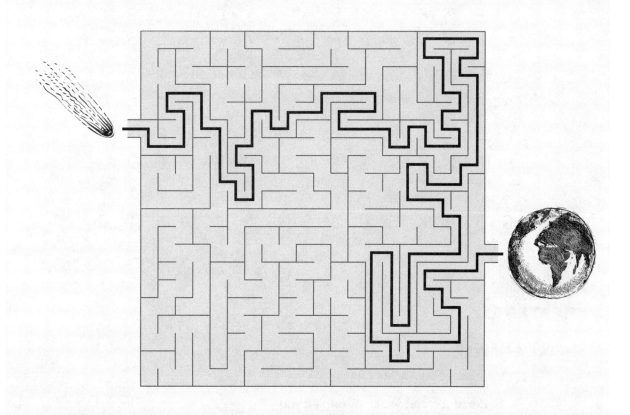

CHAPTER 10

MULTIPLE CHOICE

1.	C	4.	C
2.	A	5.	A
3.	B		

TELESCOPES WORD SEARCH

CHAPTER 11

SHORT ANSWER

1. What does NASA stand for?

National Air and Space Administration.

2. What was the "Space Race"?

The Space Race was the competition between two global powers, the United States and the USSR (Soviet Union), in the 1950s and '60s to try to outdo each other with spaceflight technology and capability.

3. What was Project Apollo? What was its primary goal?

Project Apollo was an effort by NASA to put astronauts on the Moon.

4. Who were the first two astronauts to land on the Moon?

Neil Armstrong and Buzz Aldrin.

5. What are "unmanned missions," and why are they beneficial to our exploration of outer space?

Unmanned missions are voyages into outer space with no human being aboard the spacecraft. They help us explore space because human beings are limited in how far they can travel or how fast they can go, due to biological concerns. Probes (unmanned spacecraft) can spend decades in outer space and travel great distances at great speeds.

CHAPTER 12

DEFINE THE TERMS

Exoplanet: **Any planet we observe outside our solar system (orbiting a different star).**

Star Systems: **Groups of stars that are close enough to each other to affect one another with their gravity (they all orbit each other, in effect).**

Galaxy: **A large system of stars gravitationally bound to one another.**

Black hole: **A location in space with immense mass, which generates intense gravitational forces. Gravity is so strong near black holes that nothing can escape from it, not even light.**

Light-year: **A unit of measure for long distances in space, equal to the distance light can travel in a year, an astonishing six trillion miles (~5.88 trillion, to be more precise).**

IMAGE CREDITS

Front cover Andromeda Galaxy, Tragoolchitr Jittasaiyapan / Mars, Stockbym / Saturn, Vadim Sadovski / Venus, NASA images / Asteroids, Ziben / Satellite, yoojiwhan / Reaching hands from The Creation of Adam of Michelangelo, Freeda Michaux © Shutterstock.com

pVI Sand Dollar © Mega Pixel, Shutterstock.com

pX-1 The dark night sky near Sedona, Arizona displays the wonder and beauty of The Milky Way Galaxy above Courthouse Butte and Bell Rock. © Kenneth Keifer, Shutterstock.com

p8-9 Ptolemy (c.90-168) (Claudius Ptolemaeus) in the Observatory at Alexandria, from 'La Vie des Savants Illustres' by Louis Figuier (engraving) / French School, (19th century) / Bibliotheque Nationale, Paris, France / © Archives Charmet / Bridgeman Images

p13, 110 Constellation drawings © Valentina Kalashnikova, Shutterstock.com

p16-17 Our earth in cosmos and bright sun. Elements of this image furnished by NASA © Skylines, Shutterstock.com

p23 Flying angel illustration © dmitroscope, Shutterstock.com

p23 Set of stars © The_Believer_art, Shutterstock.com

p24 Mercury © NASA images, Shutterstock.com

p24-25 Stars in space © Paolo Sartorio, Shutterstock.com

p24-25 Venus © NASA images, Shutterstock.com

p30-31 Earth and Moon in space. Elements of this image furnished by NASA © Dima Zel, Shutterstock.com

p35, 115 Moon phases, elements of this image are provided by NASA © Delpixel, Shutterstock.com

p36, 116 Crossword Puzzle © crosswordlabs.com

p38-39 Mars and its moons. Elements of this image furnished by NASA. © Vadim Sadovski, Shutterstock.com

p46-47 Conjunction of jupiter and saturn in aquarius. Elements of this image furnished by NASA. © buradaki, Shutterstock.com

p53 Jupiter and Saturn illustrations © Tartila, Shutterstock.com

p54-55 Sun © Irkin, Shutterstock.com

p58, 121 Illustration of solar system showing planets and the sun © helenpyzhova, Shutterstock.com

p59, 122 Crossword Puzzle © crosswordlabs.com

p62-63 Asteroids or meteorites field in the outer space, formation of planets. 3d illustration © Jurik Peter, Shutterstock.com

p69, 123 Halley's Comet illustration © Melok, Shutterstock.com

p69, 123 Earth illustration © Very_Very, Shutterstock.com

p70-71 Astronomer at the 15 inch Refractor Telescope © John A Davis, Shutterstock.com

p78-79 Space Shuttle Endeavour / July 17, 2009 / Photo © NASA/Novapix / Bridgeman Images

p88-89 Space background. Elements of this image furnished by NASA. © NASA images, Shutterstock.com

p102 Artist concept illustrating the relative sizes of the one-man Mercury spacecraft, the two-man Gemini spacecraft, and the three-man Apollo spacecraft. circa 1964. Source: https://images.nasa.gov/details-S64-01123 (image link); see also https://archive.org/details/S64-01123, Author: NASA/JSC [public domain] via Wikimedia Commons

p107 Full Moon © Rockline, Shutterstock.com

p107 Backside Image of Moon © Dudla Vijayanand, Shutterstock.com